AN INTRODUCTION TO
GRAND CANYON
GEOLOGY

by L. Greer Price

Grand Canyon Association
Grand Canyon, Arizona

ISBN 0-938216-68-6
Library of Congress Card Catalog Number 98-75044

10 9 8 7 6 5

Editorial: Sandra Scott
Book Design: Christina Watkins
Book Typography and Production: Triad Associates
Production Supervisor: Kim Buchheit
Editorial Assistance: Faith Marcovecchio
Printed in Singapore on recycled paper.
Cover photograph — North Kaibab Trail, Grand Canyon North Rim
Title page photograph — Vishnu Schist from Pipe Creek
Preceding photograph — Copper Canyon seen from South Bass Trail, Grand Canyon

Acknowledgments

Many thanks to Art and Peg Palmer, who reviewed the earliest draft of this manuscript. I am indebted to Stan Beus, George Billingsley, and Brad Ilg for their careful review and insightful comments. I am grateful to Carl Bowman and Jim Heywood of Grand Canyon National Park for their comments, as well. Many thanks to Mike Quinn at Grand Canyon National Park, Janice Aitchison at Lamont-Doherty Earth Observatory, George Billingsley at the U.S. Geological Survey in Flagstaff, Arizona, and Ken Cole at the Colorado Plateau Field Station of Northern Arizona University for their help in obtaining photos and illustrations. Sara Stebbins, research librarian at Grand Canyon National Park and good friend, assisted in a dozen different ways. Finally, thanks to my editor, Sandra Scott, and our designer, Christina Watkins. They are two of the best.

This book is dedicated to my advisor, mentor, and friend, Dorothy Jung Echols. It was she who first shared with me the mystery and joy of geologic science.   L.G.P.

CONTENTS

To the conception of its vast proportions must be added some notion of its intricate plan, the nobility of its architecture, its colossal buttes, its wealth of ornamentation, the splendor....

— Clarence E. Dutton

INTRODUCTION

The science of geology is one of history and process: the history of our planet, and the processes that have shaped it. It seeks to answer three very fundamental questions: What happened here? How did it happen? and How long ago? It differs from most other sciences in that it concerns itself a great deal with the past. But the issue of process — How did it happen? — is central to all science.

Geology is a mystery, a puzzle. Those who study it seek to draw as complete a picture as they can from a scant few pieces of evidence. Most of the pieces are missing, because the processes that shape our landscape are by their very nature destructive. Much of the evidence is long gone, but every now and then another piece of the puzzle turns up, allowing us to complete — or at least revise — our picture. One of the fundamental principles of geology is that the processes that operate today are the same as those that operated in the geologic past. If we can understand what's happening now, and why, we can go a long way toward solving the mystery of geologic history.

Our understanding of the grand scheme of earth science and earth history will always be incomplete. That is the lure of geology. We want to know more, there's always more to know, and the answers, we feel confident, are just around the corner. But there is beauty enough in what we know already. The essential principles of geology, simple and elegant, are expressed in the landscape we see around us — our Earth.

Grand Canyon's dramatic geology is nowhere more beautifully described than in Clarence Dutton's *Tertiary History of the Grand Cañon District,* 1882. Shown below is a portion of his map of Marble Cañon.

Nowhere is that landscape more astonishing than in northern Arizona, home to the Grand Canyon of the Colorado River. Here, in the heart of the American West, a great western river has carved its most spectacular gorge. The stretch of Colorado River that we call Grand Canyon is more than an impressive sight. In carving this gash across northern Arizona, the Colorado River has exposed millions of years of earth's history in the strata that form the canyon walls.

The process whereby such landscape is created is equally fascinating. Grand Canyon is an erosional feature. It was not carved by glaciers, nor split asunder along massive faults, but shaped by water over the course of the last few million years — simple enough, it would seem. But while the principles of erosion, like so much of geology, are simple, the detailed history of the Colorado River and its canyons remains elusive and difficult to grasp. Once on the rim, however, the issues become simpler. The questions that present themselves to all of us who stand gazing into the abyss are the fundamental questions of all geology.

Wotans Throne from Cape Royal, North Rim.

THE BIG PICTURE

The Colorado River has carved many canyons along its course, a distance of over 1,400 miles from its headwaters in the Colorado Rockies to its mouth at the Gulf of California. Its Grand Canyon stretches for 277 miles in northern Arizona. Much of the geologic history of the North American continent is revealed in the walls of Grand Canyon and in the high plateaus to its north in what has come to be known as the Grand Staircase. There are deeper canyons and wider river valleys, but nowhere has the combination

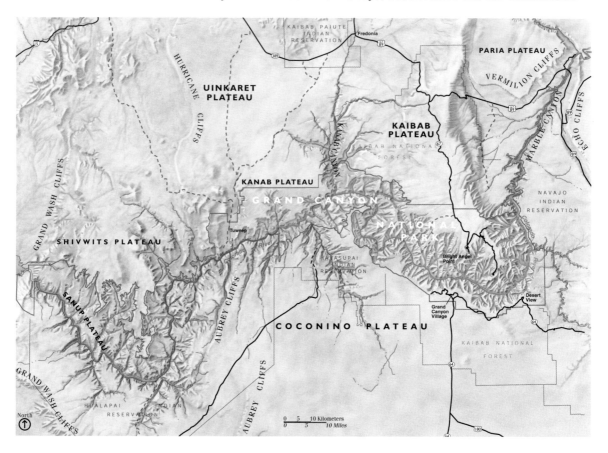

of bedrock, climate, geologic structure and history conspired to produce anything quite like Grand Canyon. It is home to hundreds of species of plants and animals, a few of them found nowhere else, and it has a rich archaeological record.

But the greatest legacy and significance of Grand Canyon lies

8

in its revelation of geologic history, and in its status as one of the finest examples of arid-land erosion in the world. While the river is the primary agent of erosion responsible for the canyon, there are other forces of erosion that continually shape its spectacular visage. Erosion is the all-encompassing term for the processes that constantly sculpt and ultimately wear down the landscape of our planet. It accompanies weathering, the ongoing break-up of materials at the earth's surface through exposure to our atmosphere and its chemical, physical and biologic processes. Erosion involves the transportation of weathered debris, from small particles of clay and sand to house-sized boulders. Gravity plays a sometimes overlooked but vital role in this process. It is urging all of that erosional debris downhill. Most of the particles are carried by running water, headed downstream to its own base level, which for nearly all water on our planet is that of the world's oceans.

The History of Geologic Exploration at Grand Canyon

The region of North America that includes Grand Canyon was one of the last areas of the continental United States to be explored and settled, owing to its remoteness, inaccessibility, and lack of water. By the middle of the nineteenth century, Grand Canyon had been visited by only a handful of Europeans, and was essentially known only to the American Indians who inhabited the region.

Between 1867 and 1879, several landmark geographical and geological surveys explored the American West. The U.S. Geological and Geographical Survey of the Territories, led by Ferdinand Hayden, was instrumental in the creation of Yellowstone National Park, and it was men attached to the Hayden Survey who were the first Europeans to see the prehistoric Puebloan cliff dwellings at Mesa Verde.

In 1869 John Wesley Powell, the ambitious explorer and geologist who led the U.S. Geographical and Geological Survey of the Rocky Mountain Region, traversed the length of Grand Canyon along the Colorado River by boat, and he is responsible for popularizing the name Grand Canyon. Though he was almost certainly the first man to complete such a trip,

Powell expedition at the mouth of the Little Colorado River.

John Strong Newberry

Grove Karl Gilbert

Powell was not the first geologist to ponder the origin and significance of Grand Canyon. Twelve years earlier, in 1857, the U.S. Army sent an expedition to the mouth of the Colorado River at the Gulf of California, under the direction of Lt. Joseph Christmas Ives. Ives' appointed task was to journey upstream as far as possible by steamship, which he did, abandoning his vessel, the *Explorer*, early in 1858 near the present site of Hoover Dam. Then on foot, he reached the western Grand Canyon somewhere in the vicinity of Diamond Creek.

Accompanying Ives on the expedition was John Strong Newberry, one of the pioneers of North American geology. It was Newberry who first recognized the canyon as "the most splendid exposure of stratified rocks . . . in the world." He drew the first geologic column, in which he recognized and named many of the individual rock layers, and noted the all-important fact that the canyon and its surrounding landscape were the result of erosion on a grand scale, ". . . wholly due to the action of water."

One of Newberry's protégés, Grove Karl Gilbert, accompanied the U.S. Geographical Survey West of the Hundredth Meridian, led by George Wheeler in 1871–1872. One of the most respected and well-known geologists of his time, Gilbert named both the Colorado Plateau and the Basin and Range provinces. He continued the work begun by Newberry, and was responsible for, among other things, naming the Redwall Limestone, the prominent red limestone cliff visible midway down the canyon wall.

In the late 1870s, Clarence Edward Dutton added his name to the list of geologists who would leave their mark on the Grand Canyon region. While best known for his studies of the landscape of the Colorado Plateau, his *Tertiary History of the Grand Cañon District*, illustrated by William Henry Holmes and published in 1881, remains one of the most beautifully written pieces of all Grand Canyon literature. Dutton, too, understood the paramount role of erosion in the creation of this landscape.

By the turn of the century, a fundamental understanding of the origin and significance of Grand Canyon had been reached. The canyon was firmly fixed not only in the minds of American geologists, but in the minds of the American public as well. Tourism at Grand Canyon, begun in earnest only in the late 1880s, caught on very quickly. By 1901 it was possible to travel by train directly to the South Rim. Today the canyon attracts millions of visitors each year (among them a healthy number of geologists), and one of the most frequently asked questions is Why here, and nowhere else? The answer is rooted in the geologic setting of this portion of North America.

Plate Tectonics

Our understanding of the origin and nature of the North American continent has come about in light of what we've learned in the past thirty years or so about the structure of our planet and the

North America is divided into a series of geologic provinces, each with certain distinct structural and physiographic features in common and sharing a similar geologic history. Each holds a unique place in the framework of the North American continent.

Grand Canyon runs across the southwest portion of the geologic province known as the Colorado Plateau. Centered roughly on the Four Corners region, where the states of Utah, Colorado, New Mexico, and Arizona meet at a common juncture, the plateau is home to some of the most spectacular scenery in America. It is characterized by thick sequences of flat-lying sedimentary rocks that are exposed in a series of high plateaus. The earth's crust is thick in this part of the country, and much of the plateau sits a mile or more above sea level.

The Colorado Plateau is fundamentally different from its neighbors. The plateau's rocks are relatively undeformed, that is to say they have not been bent or jumbled by geologic forces. This is true in spite of the fact that the Colorado Plateau has been raised thousands of feet above sea level.

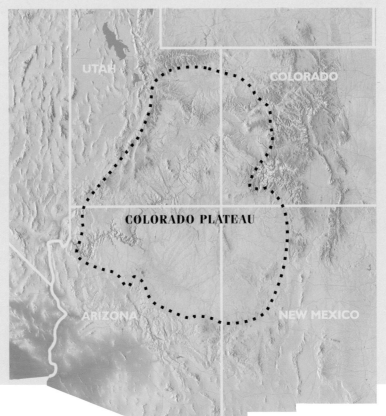

dynamic forces that shape it. Since the early part of the twentieth century scientists have known that the earth is an old planet, 4.6 billion (or 4,600,000,000) years old by current measure. In that period of time, the planet has seen a great deal of change: the origin of the continents, oceans and atmosphere, and the diverse history of

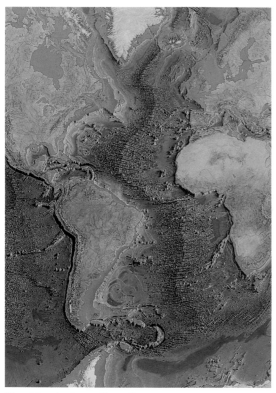

The Atlantic Ocean floor, Bruce Heezen and Marie Tharp, 1977. The Mid-Atlantic Ridge is the classic example of a divergent plate margin.

life known to us through the fossil record. It is only in the past thirty-five years that we've come to understand how the face of our planet took its present shape, and to recognize the framework upon which the continents and oceans rest.

Some of the big issues which have long fascinated us include the origin of mountains, worldwide changes in sea level, global climatic changes, the origin of ocean basins, and the seemingly random distribution of volcanoes and earthquakes. What causes huge portions of the earth's crust to rise thousands of feet above sea level? Why do we find the fossil remains of tropical plants and animals in places that today are temperate or even cold? What causes the oceans to move inland across large portions of the continents and then withdraw?

In 1912, in an effort to address some of these very big issues, a young meteorologist named Alfred Wegener proposed that the continents had migrated great distances across the face of the earth throughout geologic time. He based his theory largely on circumstantial evidence, the apparent jigsaw-puzzle fit of the existing continents, for example, and the location of nearly identical bodies of rock in widely scattered locations. But neither he nor the geologic community at large had any understanding of a mechanism that might allow such movement, and Wegener's theory of continental drift was largely discredited for decades.

The solution came following the Second World War, thanks at least partly to a renewed interest in the sea floor, a place that was as remote and unknown as the dark side of the moon. With the mapping of the sea floor came the discovery of a whole series of prominent geologic features, and the origin of continents and oceans again began to receive a great deal of attention. The discovery of mid-ocean ridges whose summits were higher than most continental peaks, and deep-ocean trenches that lay more than 30,000 feet beneath the ocean's surface, sparked a revolution in geology in the 1960s.

In the late 1950s and early 1960s, earth scientists undertook to drill through the earth's crust to learn more about the nature of our planet. Project Mohole, as it was known, received a great deal of attention, but was not in itself a great success. Penetrating the earth's crust turned out to be a difficult proposition, and not everyone was convinced that a single deep hole would answer enough questions. However, it spawned one of the most successful cooperative scientific endeavors of all time, the Deep Sea Drilling Project. The project was initiated in 1966, funded by the National Science Foundation, and operated by Scripps Oceanographic Institution under the auspices of Joint Oceanographic Institutions for Deep Earth Sampling (JOIDES), a consortium of universities and oceanographic institutions. In 1968 they launched the Glomar *Challenger,* a sophisticated research vessel capable of drilling and retrieving cores from the deep ocean floor. Over the course of the next fifteen

When I was a young graduate student, my advisor told me that there had been three great Maries throughout history: Marie Antoinette, Marie Curie, and Marie Tharp. The first two were known to me, but only a handful of geoscientists knew who Marie Tharp was. Those who did know her respected her, for it was her pioneering work at Lamont Observatory in the early 1950s (with Bruce Heezen) that produced the first map of the ocean floor. Their discovery of the series of mid-ocean ridges that circled the globe led indirectly to our understanding of sea floor spreading and plate tectonics.

The Glomar *Challenger*

13

PLATE BOUNDARIES

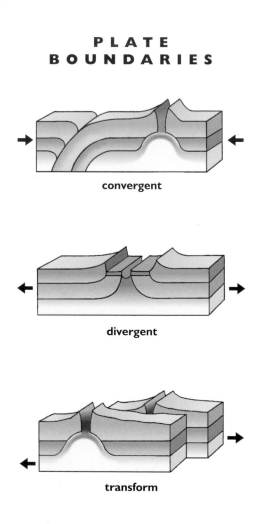

convergent

divergent

transform

years, the Glomar *Challenger* extracted thousands of cores, and subsequent discoveries associated with research on this deep-ocean material provided the answers to many crucial questions.

After years of dedicated research in many scientific disciplines, we learned that our planet's surface is divided into a series of rigid but mobile lithospheric plates, as they came to be called, that move across the face of the globe on a semi-molten interior, bearing both the continents and the ocean basins with them as they go. It became apparent that the surface of our planet was far more mobile than we'd ever guessed.

Much geologic activity is concentrated along the boundaries of these lithospheric plates. Volcanic eruptions, earthquakes, mountain building, even the rise and fall of sea level on a worldwide scale can all be related to this process of plate tectonics. What had been a dark and difficult mystery for so long was suddenly clear, and the solution to one mystery shed light on our understanding of nearly every other aspect of geology. So it often goes in the world of science: The stakes are higher than we imagine, and we may get much more than we bargained for.

We also learned that continental and oceanic crusts are fundamentally different in composition and origin. Continental crust is thick, contains significant amounts of silica, or quartz, and is much lighter than the thinner, denser oceanic crust composed primarily of the igneous rock basalt. Among other things, it seemed that, although the continents were old, the oceans were young. The sea floor in places was spreading, and along the crest of the mid-ocean ridges, new oceanic crust was created, while older slabs of oceanic crust were elsewhere being subducted, drawn down into the planet's interior along deep ocean trenches and beneath overriding continents, to be remelted and recycled.

Today we recognize three basic kinds of plate boundaries. Convergent plate boundaries occur where plates, either continental or oceanic, grind head-on into or onto one another. The convergence of two continental plates results in a great thickening of continental crust and the uplift of mountain ranges like the Himalayas. Where a continental plate converges upon an oceanic plate, the denser oceanic plate dives beneath the overriding continental plate in a process known as subduction. The subduction of an oceanic plate

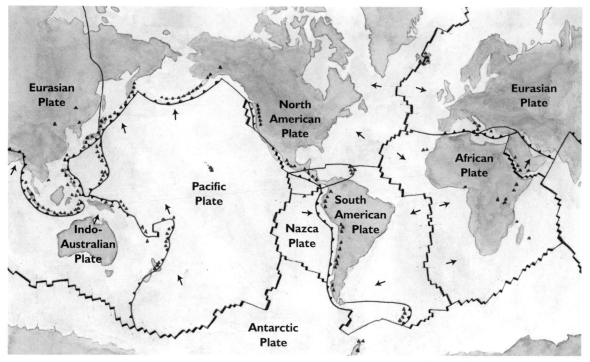

The earth's tectonic plates and their boundaries. Arrows indicate direction of plate movement; red triangles indicate zones of volcanic activity.

beneath the edge of a continent results not only in mountain build-ing and seismic activity but in volcanic activity as well, as the sub-ducted oceanic plate melts at great depth, generating molten rock which may then find its way to the surface. On the western edge of North America in Washington and Oregon, this process has pro-duced the Cascade Mountains. The eruption of Mount Saint Helens in 1980 was a very clear indication that this is an on-going process.

In the deep waters of the mid-Atlantic, new oceanic crust is created along a line that marks the boundary between the North American plate and the Eurasian plate, the classic example of a divergent plate margin. These plates are moving away from one another at a rate of two inches per year — about the rate at which your fingernails grow — and the eruption of Earth's youngest crust along this plate boundary has created the Mid-Atlantic Ridge *(see illustration on page 12)*. Ironically, its peaks are some of the highest mountains on the planet's surface, though they appear above the ocean's surface in only a few places, such as Iceland. Divergent plate margins can also occur elsewhere on the sea floor, and some-times occur mid-continent. The East African Rift, where some of the earliest remains of fossil humans have been found, is one such place. The rocks of the Grand Canyon Supergroup, a significant piece of the geologic record at Grand Canyon, may have been deposited in an older example of a mid-continent rift.

Occasionally two plates will slide by one another laterally, creating what is known as a transform plate boundary. Transform plate boundaries, unlike the other two interactions, rarely produce volcanic activity but are responsible for large-scale seismic activity. The San Andreas Fault in California is perhaps the best known example of a transform plate boundary, and the earthquakes that continue to shake the West Coast of California occur as the Pacific plate grinds northward past the North American plate. Displacement along this plate boundary in the past 20 million years has been on the order of 200 miles.

With an understanding of plate tectonics we saw the most striking physiographic features of our planet — the Himalayas, the Andes, the Cascades, and the San Andreas Fault — in a new light that clarified old questions of geologic process. We realized that even older mountains like the Appalachians were the result of continental plate collisions. Continental crust, which was old, and oceanic crust, which was not, suddenly made sense.

Of what relevance is plate tectonics to places like Grand Canyon that lie far from plate boundaries? The migration of continents through geologic time, it turns out, explains many of the global climatic changes we see in the geologic record, at Grand Canyon and elsewhere. Worldwide sea level changes can be directly related to these processes, and even the most ancient rocks at the bottom of Grand Canyon tell a story which is best understood in light of what we know today about plate movements and mountain-building processes.

The island of Surtsey, off the southern coast of Iceland, appeared in 1963 when volcanic eruptions from a divergent plate boundary (the Mid-Atlantic Ridge) reached the ocean surface.

The active volcanoes of the Cascade Range in the Pacific Northwest are the result of the subduction of the Pacific plate beneath the North American plate.

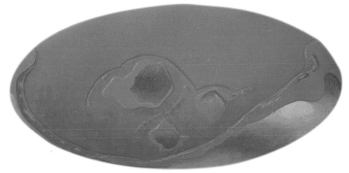

514 million years ago

306 million years ago

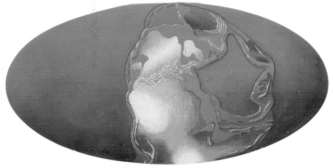

255 million years ago

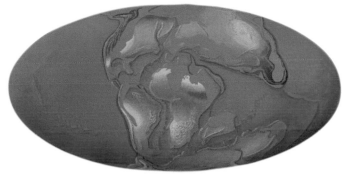

94 million years ago

Since the beginning of the Paleozoic Era nearly 600 million years ago the continents have drifted thousands of miles across the face of the globe, as demonstrated in the illustrations at left. The configuration of the continents affects many things, including the pattern of ocean currents, climate, and the distribution of plants and animals.

THE GEOLOGIC RECORD

View from the North Rim.

The Kaibab Limestone forms the canyon rim throughout the eastern Grand Canyon, including Grand Canyon Village.

We know the geologic history of our planet primarily through the rocks and landforms we see around us. There are three distinct types of rock: sedimentary, igneous, and metamorphic, all of which hold clues to their origin and history, and all of which are present in great abundance and diversity at Grand Canyon.

Sedimentary rocks are those that result from the consolidation of fragments of older rocks, carried some distance from their weathered source and deposited in layers. Sandstone and shale form from the accumulation of sand, silt, and clay, all products of erosion. These rocks can be deposited by water or wind and may come to rest in marine (ocean) or freshwater environments (lakes or streams). Limestone, composed mostly of calcium carbonate, forms through the accumulation of the secreted hard parts of marine plants and animals. Sedimentary rocks make up

only 5 percent of the rocks in the earth's crust, but they account for 75 percent of the rocks exposed on the earth's surface. Nearly all of the flat-lying rocks in the upper four thousand feet of the walls of Grand Canyon are sedimentary in origin.

The sedimentary rock record is advantageous in that many of the features typically preserved in ancient sediments — ripple marks, footprints, mudcracks, animal burrows, even raindrop impressions, and the fossils themselves — tell us a great deal about the environment that prevailed at the time and place in which they were deposited. Furthermore, these layers of sediment tend to be extensive and traceable. One of the basic tenets of geologic science is that the stuff on the bottom is older than the stuff on the top. This holds true for nearly all sedimentary rocks. (There are places where the beds have been overturned or displaced, not the case at Grand Canyon.) The correlation of sedimentary rocks of similar age and origin among far-flung locations on the globe is one of the ways in

These raindrop impressions in sandstone are evidence of a storm that occurred hundreds of millions of years ago.

Around the turn of the century scientists discovered that certain elements (like uranium) were unstable and that, through a process of natural radioactive decay, these elements changed at a predictable rate into other elements or varieties of the same element known as isotopes. This discovery led to the development of a radiometric time scale based on carefully measured decay rates of isotopes found in common rock-forming minerals. Coupled with the ability to measure isotopic ratios with some precision, we can arrive at ages for many kinds of igneous and metamorphic rocks. These "absolute" radiometric dates also make it possible to determine the ages of geologic events.

which we can unravel the history of our planet.

Igneous rocks form as molten material from the earth's interior cools and crystallizes to solid rock. When molten material erupts at the earth's surface and cools quickly to form various kinds of volcanic rock, basalt, rhyolite, obsidian, cinders, etc., it is termed extrusive. Such extrusive igneous rocks exist in remote parts of the western Grand Canyon and in the San Francisco Peaks, south of Grand Canyon near Flagstaff, Arizona. Intrusive igneous rocks, on the other hand, are those that cool slowly at great depth. These rocks may form from molten material of the same composition as that which erupts elsewhere at the earth's surface, but these intrusive igneous rocks are very different in texture. The slow cooling at great depth allows for the growth of larger crystals, visible to the naked eye. Perhaps the best known intrusive igneous rock is granite.

Granite from the Inner Gorge.

Igneous rocks rarely have anything to offer us in the way of fossils, but they hold clues for us nonetheless, for they are excellent sources of radiometric dates. These dates provide the framework upon which the entire geologic time scale is based. Without them, we would know only the relative ages of such rock units. The oldest rocks at Grand Canyon for which we have dates are igneous in origin.

Metamorphic rocks form when pre-existing igneous and sedimentary rocks are altered through a combination of heat and pressure, sometimes at great depth in the earth's crust.

Metamorphic rocks of the Inner Gorge of Grand Canyon.

Most metamorphic rocks display characteristic textures that offer information about their origins. Many are visibly contorted or tightly folded in outcrop. Metamorphic rocks like schist may display foliation, the alignment of mineral grains along planes of compression or stress, sometimes mistaken for sedimentary bedding. The most important sign of their origin is the presence of certain minerals that are associated almost exclusively with metamorphic rocks. Garnet, for example, is a mineral typically found in metamorphic rocks. Such minerals give us clues to the very specific conditions under which the rock in question was altered.

The Precambrian Rocks of Grand Canyon

The oldest rocks at Grand Canyon are the complex group of igneous and metamorphic rocks known as the Granite Gorge

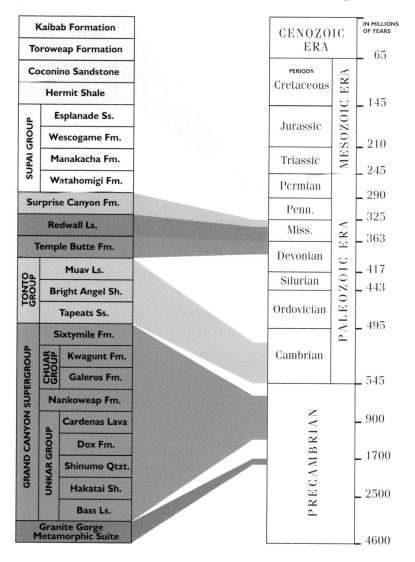

Today we use as a frame of reference a geologic time scale, developed over a period of more than a hundred years by geologists working together to correlate rock layers with similar fossils and age characteristics. This relative time scale is linked to specific radiometric dates in order to give absolute ages to both the rocks and the events which shaped them. Throughout this book I refer to specific names of periods, epochs, and eras which are tied to this scale. A quick glance at the stratigraphic section for the Grand Canyon *(on the left)* from time to time will also help put things in their proper perspective. A comparison of the section and the time scale *(on the right)* gives one an immediate notion of how little of Earth's history is represented in the geologic record, even at a place like Grand Canyon.

Metamorphic Suite, exposed at river level in the Inner Gorge. Here these rocks form the very basement of the North American continent. Such basement rocks, generally the oldest rocks on any continent, are often buried beneath thousands of feet of younger sedimentary layers, as they are at Grand Canyon and throughout much of North America. They are exposed at the surface only where erosion has removed the overlying rock, in deeply incised canyons (Grand Canyon), or in mountainous regions where uplift has facilitated the erosion of overlying sediments (the Black Hills of South Dakota, the Ozark Mountains of southeast Missouri, and the Rocky Mountains). Such rocks are also well exposed in places where glaciers have stripped away the younger, softer rocks, as in the Canadian Shield.

Overshadowed from the rim by the sedimentary rock walls above them but dominating the scene at river level, the Precambrian rocks of Grand Canyon have much to tell us. From the canyon rim west of Grand Canyon Village toward Hermits Rest, one can easily spot the dark, contorted shapes of the schists and the lighter-colored veins of granite in the Inner Gorge. These rocks appear ancient, and radiometric dates substantiate that they are as old as 1.84 billion years. They hold important clues about the growth of the North American continent, and only recently has our understanding of plate tectonics allowed us to fully appreciate their significance.

When these basement rocks were formed, the face of the globe was very different than it is today. The continents in general were smaller, and the configuration of oceans and continents was different as well. The area in which we now find these

Geologists reworking the Precambrian rocks of the Inner Gorge in recent years have renamed the entire group of basement rocks the Granite Gorge Metamorphic Suite, which includes the metamorphosed sediments we call the Vishnu Schist and the metamorphosed volcanics of the Brahma and Rama Schists. These rocks formed near the earth's surface during plate collisions. Molten materials generated at great depth during these collisions slowly rose to intrude the overlying rocks. At Grand Canyon these intrusions, or plutons, include the Ruby, Diamond Creek, Pipe Creek, and Surprise plutons.

volcanics

sediments

pluton

Precambrian rocks was then sea floor, lying between a younger, smaller, North American continent and a series of volcanic islands offshore. These volcanic islands formed above subduction zones, where one oceanic plate was sinking beneath another, a process that generates both heat and molten material.

People who are interested in the oldest rocks on the planet might well expect to look at the bottom of Grand Canyon to find them, given that the Colorado River has exposed rocks which elsewhere lie thousands of feet below the surface. While they are indeed the oldest rocks in this region and certainly of respectable age (close to 2 billion years), the oldest rocks at Grand Canyon (the Elves Chasm Gneiss, at right) are far from the oldest rocks in the world, or even North America. We find the oldest known rocks — close to 4 billion years old — on the shores of Great Slave Lake in Canada's Northwest Territory, and in remote areas of Greenland.

A collision-generated granite cross-cuts this Elves Chasm Gneiss.

About 1.7 billion years ago, these ancient volcanic islands, along with perhaps a few smaller pieces of continental crust, collided with the North American continent. During this collision the crust nearly doubled in thickness, and a vast amount of new continental crust was formed. In a relatively short time North America grew from the Wyoming–Colorado border to well south of the U.S.–Mexico border. The subducted crustal slabs melted at great depth to form bodies of molten rock (plutons and dikes) which rose through the crust, intruding the crushed volcanic islands and ocean sediments. Some of the molten rock cooled to form the pink veins of quartz that lace the walls of the Inner Gorge.

Lying above these older Precambrian rocks are remnants of younger Precambrian rocks known as the Grand Canyon Supergroup. Although present only as isolated remnants, they represent a significant portion of the geologic record at Grand Canyon. Ranging in age from about 800 million to 1.2 billion years old, they include a complex assortment of sedimentary and igneous rocks. Together their layers are around 12,000 feet thick. Though they lie directly

The rocks of the Grand Canyon Supergroup, inclined at an angle of about twenty degrees, are well-exposed and clearly visible from the rim in the vicinity of Desert View.

upon older Precambrian rocks, they were laid down after a long period of erosion had removed intervening layers. This sort of gap in the geologic record is called an unconformity. The rocks of the Grand Canyon Supergroup were deposited over a period of about 400 million years. Following their deposition, they were subjected to uplift, tilting, faulting, and erosion. These rocks may have formed in a continental rift similar to the East African rift. The Grand Canyon Supergroup rift failed to split apart the continent but did allow for the accumulation of a thick sedimentary record.

The rocks of the Grand Canyon Supergroup are best seen in the eastern Grand Canyon, from Lipan Point, Moran Point, and Desert View. Tilted at an angle of about twenty degrees, they are easily spotted beneath the flat-lying Paleozoic strata. Along the stretch of canyon visible from the road to Hermits Rest, rocks of the Grand Canyon Supergroup are missing altogether.

Constructive geologic processes — mountain building, uplift, the deposition of sediments, volcanic eruptions, etc. — are balanced by destructive processes, primarily erosion in all its varied forms, which destroy both landforms and the existing geologic record. The

The oldest fossils on Earth are microscopic organic structures found in rocks as old as 3.6 billion years. Stromatolites, layers of limy sediment trapped and bound by mats of cyanobacteria (blue-green algae) constitute the oldest fossils at Grand Canyon and are well represented in the billion-year-old Bass Limestone (at left). Their distinctive remains, easily spotted with the naked eye, leave little doubt as to their origin. Stromatolites (and a few other primitive single-celled organisms) are the only record of life we have on this planet for a period of nearly 3 billion years, and are still with us today. Their modern-day analogues (below) may be seen in Shark Bay, Australia, a shallow, marine environment similar to the one in which they flourished long ago.

early history of our planet is all but lost to us, the oldest rocks reaching back only as far as 4 of the 4.6 billion years in question. The oldest recognizable sediments are about 3.8 billion years old. What's more, the sedimentary rock record is full of gaps, places where we either have no record (because it was never deposited), or where the record was subsequently removed by erosion, a condition we term an unconformity.

As a general rule, the most significant unconformities are likely to be the easiest to spot, and that certainly is the case at Grand Canyon. The stretch of geologic time between the igneous and metamorphic rocks of the Inner Gorge and the oldest Paleozoic sediments in the canyon — a gap of over a billion years — constitutes what has long been known as the Great Unconformity. Named by John Wesley Powell in the late 1800s, it is clearly identified as the gap between the crystalline rocks of the Inner Gorge and the horizontally bedded Paleozoic rocks above.

The geologic record is fragmentary and incomplete, even at places like Grand Canyon. Within its thick sequences of rock, the intervals of time not represented are far greater than those that are. Such gaps are known as unconformities, and their presence implies a period of uplift and erosion. Sometimes it is readily apparent that layers are missing, but more often it is only through a careful examination of discontinuities in the fossil record or radiometric dates that we recognize such gaps. The **Great Unconformity** (*at left*) is visible as the line which separates the Precambrian rocks of the **Inner Gorge** from the overlying horizontal Paleozoic strata.

In fact, there are two major unconformities: Powell's Great Unconformity and the smaller gap between the gently inclined layers of the Grand Canyon Supergroup and the horizontal Paleozoic rocks directly above. Exposures of the Grand Canyon Supergroup are isolated and discontinuous. Throughout much of the canyon, the younger Paleozoic strata sit directly on top of the older crystalline basement rocks.

GEOLOGIC LAYERS OF THE GRAND CANYON

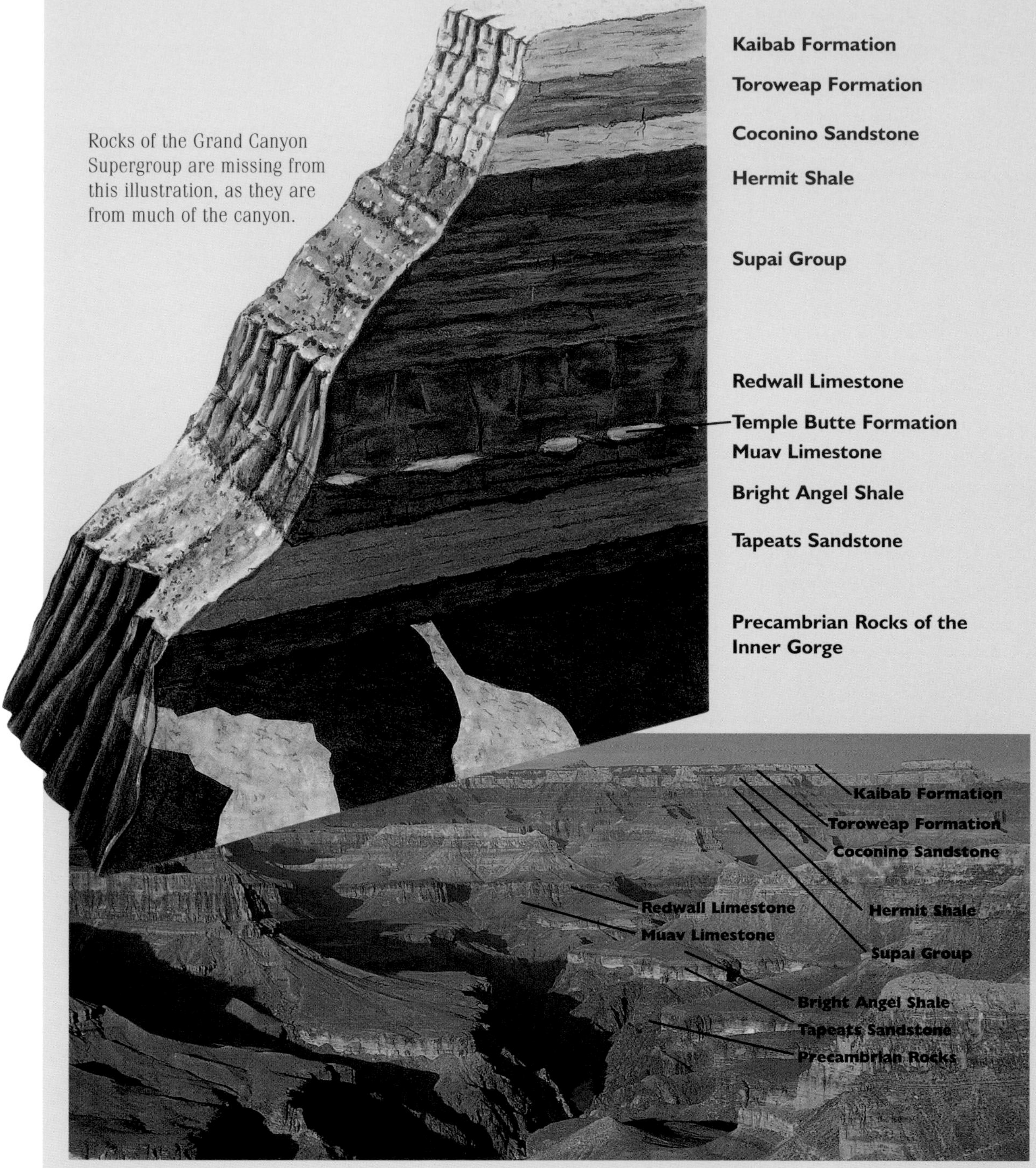
Kaibab Formation
Toroweap Formation
Coconino Sandstone
Hermit Shale
Supai Group
Redwall Limestone
Temple Butte Formation
Muav Limestone
Bright Angel Shale
Tapeats Sandstone
Precambrian Rocks of the Inner Gorge
Kaibab Formation
Toroweap Formation
Coconino Sandstone
Hermit Shale
Supai Group
Redwall Limestone
Muav Limestone
Bright Angel Shale
Tapeats Sandstone
Precambrian Rocks

Rocks of the Grand Canyon Supergroup are missing from this illustration, as they are from much of the canyon.

The Paleozoic Record at Grand Canyon

Virtually all of the horizontally bedded sedimentary rocks visible from the rim of Grand Canyon are Paleozoic in age. Ranging from early Paleozoic (about 550 million years old) to late Paleozoic (about 250 million years old) from bottom to top, these rock layers give us an incomplete but valuable glimpse of what was going on in this part of North America during that time. Rocks of Paleozoic age are scattered throughout the world, but there are few places where they are so well exposed.

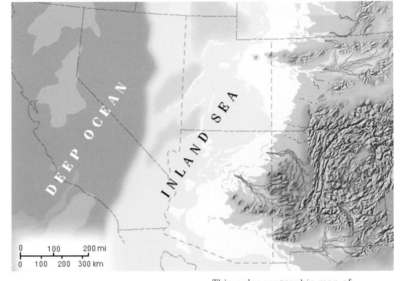

The term formation refers to any rock unit that is recognizable and traceable for a long distance. Nowhere are formations more clearly seen than at Grand Canyon, where most of the layers visible from the rim can be traced for many miles throughout the canyon. These individual rock layers may become thicker or thinner, or even disappear altogether, but each distinctive layer of sedimentary rock is distinguishable from the units above or below it, sometimes readily, but generally upon careful examination.

This paleogeographic map of western North America during the early Paleozoic indicates the position of the Cambrian ocean relative to the North America we know today. The west coast of the continent was considerably closer to the heartland than it is today.

The division of geologic time into eras and periods *(see geologic time scale on page 21)* is somewhat arbitrary, but it reflects some key events in the history of our planet. The beginning of the Paleozoic was characterized throughout the world as a time of transgression: Existing seas began to move onto the land, covering continental areas with shallow waters. This is hard for us to picture, because today the continents are high and dry and marine waters are largely confined to deep basins. We have very few shallow intercontinental seas on the planet today (Hudson Bay in Canada is one example), but these seas were a hallmark of deposition throughout much of the 300 million years of Paleozoic time. They reached a culmination during the late Ordovician, at which time they covered much of the North American continent.

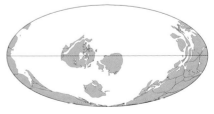

Late Cambrian

Depositional environments for the rocks of the Tonto Group. Future studies of the Bright Angel Shale are likely to reveal a richer fauna than that illustrated here. From top to bottom: Muav Limestone, Bright Angel Shale, Tapeats Sandstone.

At Grand Canyon this story of oceanic transgression in the early Paleozoic is beautifully revealed in the rocks of the **Tonto Group**, a classic transgressive sequence. The lower three Paleozoic rock units, reflecting the gradual invasion of marine waters from the west, reveal 1) a beach or shoreline environment (the **Tapeats Sandstone**), changing to 2) a quieter offshore marine environment characterized by the accumulation of muds (the **Bright Angel Shale**), then to 3) the still quieter waters and limy sediments, even farther offshore (the **Muav Limestone**).

At the beginning of the Paleozoic Era there is a "sudden" appearance of life in the fossil record. While late Precambrian seas were by no means devoid of life, during the Cambrian Period a remarkably diverse assemblage of invertebrate creatures — crustaceans of various kinds, echinoderms, and a host of others — made their rapid appearance. In western North America, this proliferation of life probably took place over a period of 15 million years or so. Little work has been done on the Cambrian fossils from Grand Canyon, but we do find trilobites and brachiopods, particularly in the Muav Limestone and Bright Angel Shale. Although this appearance of invertebrate fossils at the beginning of the Paleozoic may be due in part to the development of preservable hard parts, it seems likely that it was a time of great success and rapid evolution among invertebrates. The worldwide transgressions of the Cambrian Period provided warm, shallow seas with nutrient-rich waters in which these animals thrived.

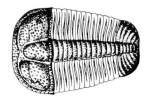
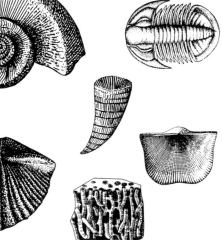
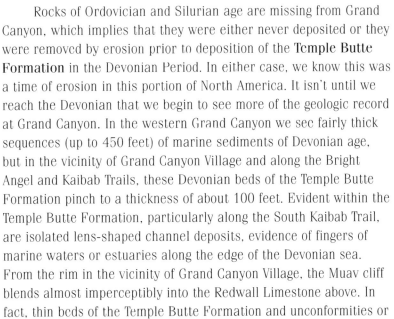

Invertebrate fossils from the Paleozoic rocks of Grand Canyon.

Rocks of Ordovician and Silurian age are missing from Grand Canyon, which implies that they were either never deposited or they were removed by erosion prior to deposition of the **Temple Butte Formation** in the Devonian Period. In either case, we know this was a time of erosion in this portion of North America. It isn't until we reach the Devonian that we begin to see more of the geologic record at Grand Canyon. In the western Grand Canyon we see fairly thick sequences (up to 450 feet) of marine sediments of Devonian age, but in the vicinity of Grand Canyon Village and along the Bright Angel and Kaibab Trails, these Devonian beds of the Temple Butte Formation pinch to a thickness of about 100 feet. Evident within the Temple Butte Formation, particularly along the South Kaibab Trail, are isolated lens-shaped channel deposits, evidence of fingers of marine waters or estuaries along the edge of the Devonian sea. From the rim in the vicinity of Grand Canyon Village, the Muav cliff blends almost imperceptibly into the Redwall Limestone above. In fact, thin beds of the Temple Butte Formation and unconformities or gaps representing about 150 million years separate the Muav and Redwall Limestones.

At least to some degree, the great diversity of sedimentary rocks we see at Grand Canyon is the result of rising and falling sea levels throughout the 300 million years of the Paleozoic. We recognize two kinds of sea level changes: those that are worldwide (in which case one might look for evidence of such transgressions in rocks of similar age around the planet), and those that are more local and likely the result of a rising or falling landmass, rather than any worldwide event. What would cause a change in sea level around the world? One obvious factor is the increase or decrease in size of the polar ice caps. If a great deal of water is locked up in permanent bodies of ice, then the volume of water in the ocean basins will drop. One concern of global warming is that a melting polar ice cap would flood the coastal cities of the world. The sea level changes we see throughout the Paleozoic, however, are more than likely related not to a changing volume of water, but rather to an increase or decrease in the volume of the ocean basins themselves because of the fluctuating rate of sea-floor spreading.

The Late Paleozoic at Grand Canyon

Depositional environment of the Redwall Limestone.

A period of erosion lasting 40 to 50 million years separates the Temple Butte Formation from the overlying Mississippian Redwall Limestone. The **Redwall Limestone** is easily recognized as the broad, cliff-forming unit whose red-stained face is prominent throughout the canyon. Hikers know it well, for the Redwall cliff is often the major obstacle in hiking from rim to river. In areas where trails are present, a long series of steep switchbacks provides access through the 500-foot cliff.

During the Mississippian Period marine waters once again covered large portions of the continent. At the time, North America lay closer to the equator, and the shallow, warm, and well-lighted waters of the Redwall sea were teeming with invertebrate faunas. Crinoids, corals, bryozoans, brachiopods, and cephalopods are all well represented in the fossil record. The period of time represented by this thick accumulation of limestone is probably in the neighborhood of 40 million years. It was, in general, a time of tranquility in North America.

Sandwiched between the Redwall Limestone and the Supai Group is a rock unit known as the **Surprise Canyon Formation**. It consists of small patches or lenses of marine sediment deposited in channels of an ancient stream system that formed on the eroded surface of the Redwall Limestone. The Surprise Canyon Formation is well developed and well exposed in the western Grand Canyon, but less so in the eastern part near Desert View or even Grand Canyon Village. It contains fossils of both plants and marine invertebrates, indicating that the stream system was flooded by an advancing sea which transformed it into a series of estuaries. Like the Temple Butte Formation, it is virtually invisible from the rim in the vicinity of Grand Canyon Village. Its recent discovery, in 1973, is a reminder that Grand Canyon has yet to reveal all of its secrets.

Pennsylvanian

Depositional environment of the Supai Group.

Above the Redwall Limestone are the strata of the **Supai Group**, nearly 1,000 feet of red beds that form a series of cliffs and slopes between the Redwall Limestone and the Hermit Shale. Separated from the Redwall Limestone by an unconformity of 15 million to 20 million years, the strata of the Supai Group range in age from 270 million to 320 million years. These sediments began to accumulate toward the end of the Mississippian Period, continued throughout all of the Pennsylvanian and into the Permian. Shale, siltstone, sandstone, and limestone are represented here, divided into four individual formations that are difficult to distinguish from the rim. The wide variety of rock types present within the Supai Group, representing both marine and non-marine environments, reflect rapidly changing coastlines and rising and falling sea levels. In general we can say that the Supai Group was deposited on a broad coastal plain, not unlike areas of the Texas Gulf Coast today.

Depositional environment of the Hermit Shale.

Directly above the Supai are the four remaining units of the Paleozoic Era, all Permian in age. The Permian was in general a time of regression; the seas gradually retreated to the deeper ocean basins. The climate of western North America was warm and dry. The **Hermit Shale**, more properly known as the Hermit Formation, is composed mostly of silt, mud, and fine-grained sand and was probably deposited by streams on a broad coastal plain, much like the Supai Group beneath it but perhaps reflecting the increasingly terrestrial nature of the environment. Plant fossils in the Hermit Formation support this interpretation. At least some period of erosion separates the Hermit Formation from the overlying Coconino Sandstone, for in the top of the Hermit we find deep mudcracks filled with Coconino sands.

The **Coconino Sandstone** is easily recognized as the broad, light-colored band of rock that lies only a few hundred feet below the rim. Along the Bright Angel and the North and South Kaibab Trails the Coconino can be closely examined. Within the formation itself, the bedding planes run at steep angles to one another, a distinctive feature known as cross-bedding. This kind of high-angle cross-bedding is typical of sandstones that are deposited not by water, but by wind, and represent the faces of ancient sand dunes, cemented in place and frozen in time. The absence of silt and clay, the uniform size of the sand grains, and the absence of marine fossils further support this interpretation. Finally, the presence of distinctive invertebrate and vertebrate trace fossils, which include the fossil footprints of land-dwelling reptiles (older and smaller than the dinosaurs of the Mesozoic Era still to come) are convincing evidence of the terrestrial origin of the Coconino Sandstone. The sandy Coconino desert prevailed for perhaps 5 million to 10 million years,

Cross bedding is evident in the Coconino Sandstone, even at a distance. Note the inclined internal bedding surfaces within the horizontal upper and lower boundaries of the formation.

Depositional environment of the Coconino Sandstone.

further testament to the warm, arid, terrestrial environments which prevailed in this part of North America at the end of the Paleozoic. Seas returned to the Grand Canyon region, and in the eastern Grand Canyon the sands of the Coconino represent coastal dunes that lay to the east of the advancing Toroweap sea.

The **Toroweap** and **Kaibab Formations**, the two rock units that complete the geologic section at Grand Canyon, contain marine limestones and shales. The presence of gypsum in both of these formations is an indication of the occasionally shallow and restricted nature of the marine environment that prevailed. At its height, the Kaibab sea advanced to the east as far as the present-day Little Colorado River. The limestones of the Kaibab Formation, which form the rim of the canyon for much of its

Reptile tracks in the Coconino Sandstone.

Depositional environment of the Toroweap and Kaibab Formations.

length, comprise the single rock unit most visitors are likely to see up close. The marine fossils it contains provide clues to its origin.

The End of the Paleozoic Era

Throughout the late Paleozoic, the North American continent was headed toward a collision with a large continental landmass on its eastern margin. This collision was in many ways the culminating event of the late Paleozoic. The Appalachian Mountains are one result of this collision, as continental crust was welded onto the eastern edge of the continent. The Appalachians of today, which roughly define the line along which this collision occurred, are just a shadow of their former selves, having been subjected to erosion for the past 200 million years. In their time, they must have been imposing.

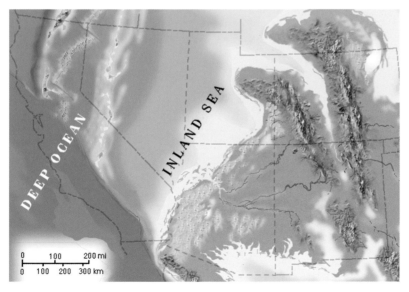

Paleogeographic map of western North America during the Permian.

By the end of the Paleozoic, 250 million years ago, all of the major continents had collided with one another to form one large continental landmass known as Pangea. The Atlantic Ocean did not yet exist, and would not begin to open for some time yet. Finally, the end of the Paleozoic was a time of regression, and it is here that the Paleozoic record at Grand Canyon (and throughout much of the world) comes to an end.

The end of the Paleozoic was also marked by the greatest extinction evident in the fossil record. Up to 90 percent of the marine organisms that flourished in the Paleozoic seas (including trilobites) were gone by the beginning of the Mesozoic Era. This Permian extinction seems to have affected sea-dwelling organisms more than those on land, and evidently was not as sudden as the one still to come at the end of the Mesozoic, which marked the extinction of the dinosaurs. While much light has recently been shed on the Mesozoic extinction, the cause of the Permian extinction is more difficult to pin down. It may have resulted from cumulative factors, including the global deterioration of marine environments. Whatever the cause, many of the species that graced the fossil record throughout the Paleozoic, at Grand Canyon and elsewhere, are seen no more.

Permian

Fossils of the Kaibab Formation. The small, disk-shaped fragments are remains of crinoids, primitive echino-derms that flourished in the shallow inland seas of the late Paleozoic. They are among the most common fossils seen at Grand Canyon.

The fossil record at Grand Canyon is rich, from the stromatolites of the Bass Limestone to the Pleistocene cave remains. The visitor with only a few hours to spend at the canyon is most likely to see the marine invertebrate fossils in limestones of the Kaibab Formation, visible along the rim where erosion has highlighted them, or along the upper reaches of the South Kaibab Trail. There are brachiopods, sponges, and crinoids, although many of these remains are broken or fragmented. For those who venture some distance below the rim, the most easily recognized (but still elusive) fossils are the reptile footprints in the Coconino Sandstone.

The Hermit Shale has a wealth of plant remains, but they are rarely seen in the more heavily visited areas of the canyon. There are also animal tracks in the rocks of the Supai Group. The business of detecting fossils in the rocks of Grand Canyon is a tricky one; consider yourself lucky if you come across one.

The thrill of finding fossil remains of an animal that lived millions of years ago is the reward for those with keen eyes and determination. Please leave them as you find them so others can experience that same thrill. All fossil remains within the park are protected by law. Collecting is strictly prohibited. The same holds true for all natural and historic objects; the removal of even one small piece of rock is a violation of federal law.

The Mesozoic Era at Grand Canyon

The stratigraphic record at Grand Canyon comes to an abrupt halt at the end of the Paleozoic Era. Only in small isolated patches on the rim of Grand Canyon are there rocks younger than the Kaibab Formation. This reflects to some degree the history of the North American continent. At the end of the Paleozoic most of the shallow marine waters that had covered so much of the continent retreated. By the beginning of the Mesozoic, much of North America was high and dry, and the record of Mesozoic rocks is largely one of terrestrial deposits: river mud flats, sand dunes, etc. During this time great thicknesses of windblown sands, like the Navajo Sandstone, accumulated throughout the West, including the Grand Canyon region. Conservative estimates are that somewhere in the neighborhood of four or five thousand feet of Mesozoic sediment accumulated in the Grand Canyon region, but whatever was there has long since been removed by erosion, following the later uplift of the Colorado Plateau. Only isolated remnants of Mesozoic rocks are present nearby, most notably at Cedar Mountain, just east of Desert

View, and at Red Butte, a few miles south of Grand Canyon.

What we do know about the Mesozoic comes to us from extraordinary exposures of rock in far northern Arizona and southern Utah. Here, in a series of high plateaus known as the Grand Staircase, thousands of feet of Mesozoic rock are preserved and exposed at places like Zion National Park, the Vermilion Cliffs, Glen Canyon National Recreation Area, and Grand Staircase–Escalante National Monument. Triassic rocks on the Navajo Reservation to the east of Grand Canyon contain footprints of large reptiles. Jurassic and Cretaceous rocks may be found on Black Mesa in northeastern Arizona, at Mesa Verde in southwestern Colorado, and in Utah. Perhaps the best-known Mesozoic rocks of the region are the brightly colored shales of the Triassic Chinle Formation which make up that remarkable landscape to the east of Grand Canyon known as the Painted Desert. These shales constitute the bulk of the landscape at Petrified Forest National Park and include abundant petrified wood.

Mesozoic rocks are spectacularly exposed in the Vermilion Cliffs along Highway 89A in northern Arizona.

These were interesting times. It was during the Mesozoic that birds and early mammals appeared, as did the bees, and flowering plants. This was the era dominated by the dinosaurs. During the 180 million years of the Mesozoic Era, the supercontinent of Pangea began to break apart, and the Atlantic Ocean was born. By the end of the Mesozoic, the face of the globe resembled our modern world a great deal more than it did at the end of the Paleozoic.

The familiar catastrophic extinction at the end of the Mesozoic Era 65 million years ago was the end not only for the dinosaurs but for perhaps half of the species in existence at the time. Large and small, plants and animals, marine and terrestrial, vertebrates and invertebrates all were affected. And while the extinction at the end of the Paleozoic was more than likely gradual (at least by our standards), the extinction at the end of the Mesozoic was evidently far more abrupt and can be traced to a single catastrophic event. Exposures of Cretaceous–Tertiary boundary rocks in New Mexico are those closest to Grand Canyon. There, too, in rocks that

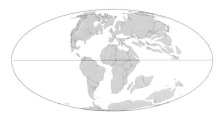

Late Mesozoic

Many geologists now believe that the impact of an asteroid or comet was largely responsible for the massive extinctions at the end of the Mesozoic.

represent the very end of the Mesozoic, is evidence of this world-wide catastrophe.

We now believe that the end of the Mesozoic coincided with the collision of a large comet or asteroid with our planet in the vicinity of Mexico's Yucatan peninsula. The energy released during this impact has been estimated at 10,000 times that of the entire nuclear arsenal of the world today. Shock waves, tidal waves, an immense fireball, forest fires of global proportions and a period of intense heat, followed by a period of cold and darkness that may have lasted several years, all contributed to the great dying off that followed.

The Laramide Orogeny

Orogenies are times of uplift, volcanic activity, thrusting, folding, faulting, and intrusion of igneous rocks (like granite) which can be radiometrically dated and therefore help us determine the timing of such events. The Laramide Orogeny, that mountain-building event that began at the end of the Mesozoic, continued into the Eocene, and culminated in the uplift of the Rocky Mountains, was no exception. The San Rafael Swell, the Monument Upwarp (now the site of Monument Valley), the Kaibab Uplift, and the Uinta Mountains in northern Utah all formed or became more pronounced at this time.

The folds, faults, and uplifts — all prominent structural features of the Grand Canyon region — are clear indications of the late Mesozoic restlessness of the earth's crust throughout this part of the country.

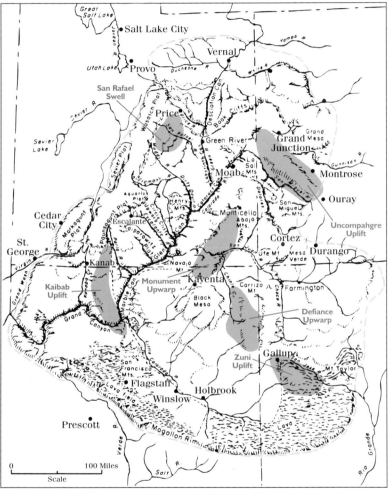

Major uplifts on the Colorado Plateau.

We can look to plate tectonics for an explanation of this upheaval. Far to the west, off the coast of a somewhat smaller North American continent, the Pacific plate was being subducted under the North American plate, as it is today off the coast of Oregon and Washington. Compression and the presence of the subducted plate beneath the continent created a thick crust. By the end of the Laramide Orogeny, the basic structural features of the Colorado Plateau — and the region that would later be home to Grand Canyon — were established.

It is important to remember that at least some of the uplift that took place here occurred much later. Perhaps as much as several thousand feet of regional uplift throughout the Colorado Plateau has occurred in the past 5 million years. It is this most recent pulse of uplift that has allowed all of the rivers on the Colorado Plateau, from the San Juan to the Paria, to carve the deeply incised canyons for which the plateau is known. And this recent uplift, along with accompanying climatic changes, has helped shape the arid lands of the Colorado Plateau. Although the framework of the Colorado Plateau is quite ancient, the face of this land and the canyons for which it is known are quite young.

The Cenozoic at Grand Canyon

The Cenozoic includes the period of time from the end of the Mesozoic to the present. In that 65-million-year period, many significant events occurred: the appearance and extinction of entire populations of immense, bizarre-looking mammals that wandered throughout North America; the evolution of primates (including that strange creature we call *Homo sapiens*); the evolution of the horse; and, toward the end of it all, the carving of Grand Canyon.

Unfortunately, the geologic record for the Cenozoic Era at Grand Canyon is extremely limited. Extensive sedimentary layers, like those of Paleozoic age that we see in the canyon walls, are virtually absent. There are good reasons for this. Thanks to the Laramide Orogeny at the end of the Mesozoic, this part of the country was high and dry for most of the Cenozoic. The predominant geologic force at work was erosion. It was during the Cenozoic that thousands of feet of Mesozoic strata were stripped away. There are, however, a few deposits from this time period that are of great interest and significance. They give us a glimpse, at any rate, of what was happening here toward the end of the Cenozoic Era.

Cenozoic Volcanic Activity

When John Wesley Powell reached the central part of Grand Canyon during his pioneering trip through the canyon in 1869, he was astonished to see cascades of lava frozen to the canyon walls. "What a conflict of water and fire there must have been here!" he wrote. "Just imagine a river of molten rock running down into a river of melted snow. What a seething and boiling of the waters; what clouds of steam rolled into the heavens!"

Volcanic activity in the past 3 million years produced a series of lava flows and associated volcanic features from southwest Utah to the rim of Grand Canyon. The small range of mountains known as the Uinkarets, visible on the distant western horizon from points along the road to Hermits Rest, was created as a result of this activity. Volcanic eruptions formed cinder cones on both sides of the canyon and below the rim. Rivers of lava cascaded over the canyon rim into the river below, not once but many times. This occurred far to the west of Grand Canyon Village in a remote area of the park seen today by few visitors, near the foot of the Toroweap Valley.

The lava cascades that Powell identified effectively dammed the Colorado River for hundreds of years. These dams, the highest of which would have towered above Glen Canyon Dam, created lakes that extended upstream as far as present-day Moab, Utah. Remnants of individual lava dams have been identified and are evident along the banks of the Colorado River between river miles 177 and 190. Lake sediments from the reservoirs created by these dams have also been found in a few isolated places upstream. The fertile farmland occupied by the Havasupai Indians along Havasu Creek is likely a delta formed by one of these lakes.

Remnants of the lava cascades at Toroweap, in the western Grand Canyon. Vulcans Throne, the cinder cone perched on the rim, is one of many that erupted in the vicinity.

Cenozoic volcanic activity is evident to the south of the canyon, as well. The San Francisco Peaks, Bill Williams Mountain, and Sunset Crater (near Flagstaff) are all part of the San Francisco Volcanic Field that has been active, off and on, during the past 6 million years. Sunset Crater, which erupted in A.D. 1064, is pretty convincing evidence of our planet's ongoing geologic activity.

The Pleistocene Epoch

Toward the end of the Cenozoic, much of the earth's climate underwent a period of cooling sometimes referred to as the Ice Age but more properly known as the Pleistocene Epoch. Scientists traditionally have recognized four distinct periods of cooling during the Pleistocene but now identify many more than that. The first of these began a little less than 2 million years ago. During these periods, the general climate of the earth cooled to such a degree that glaciers expanded in size and extent. Continental glaciers, usually confined to the polar regions, extended into territory we consider

Horseshoe Mesa as it might have appeared in the late Pleistocene, about 20,000 years ago *(above, in a computer-enhanced image)* and Horseshoe Mesa today *(below)*. During the coolest part of the late Pleistocene, most plant species grew at elevations 2,500 feet lower than today. The upper half of the canyon contained a coniferous forest of spruce and fir, while the lower half supported a cold and dry juniper woodland similar to many parts of Utah today.

temperate today. Glaciers covered most of the high peaks of North America, including the San Francisco Peaks in northern Arizona. These glacial periods were separated by intervals of warming that are known as interglacials.

During the most extensive period of glaciation in North America, continental ice sheets covered most of Canada and portions of the northern United States, reaching as far south as Missouri. While the upper reaches of the Colorado River system must have been impacted by these events, Grand Canyon itself was untouched by glaciers. We know this with some certainty, because glacial valleys (such as Yosemite, in the Sierras of California) have a unique shape, quite different from the erosional face of Grand Canyon.

The Grand Canyon region was affected by the cooling trends that accompanied the advance of glaciers in the north. River flows responded to glacial melting far upstream. Vegetation patterns were entirely different during the coldest periods. The evergreen forests, today limited to elevations above 7,000 feet, extended well into the canyon. Plants and animals that inhabited the canyon during the Pleistocene were very different from what we see today. We know something about this, thanks to an unusual discovery that was made at Grand Canyon in the 1930s.

The Evidence From Rampart Cave

At river mile 275, in the western Grand Canyon, there is a small cave at the base of the Muav Limestone, about 700 feet above the river. Here in 1936 a National Park Service employee named Willis Evans made one of the most significant paleontological discoveries in this part of the world. In what today is known as Rampart Cave, Mr. Evans discovered Pleistocene deposits that included hair, dung, claws, bones, and desiccated parts of internal organs of *Nothrotheriops shastense,* the Shasta ground sloth. There were also the remains of an extinct horse, camel, and mountain goat. (The Spanish reintroduced the horse in this region in the sixteenth century.) The remains were exceptionally well preserved. Discoverers remarked at the time that they were still quite fragrant.

These Ice-Age herbivores evidently inhabited the cave intermittently for a period of close to 30,000 years, from around 40,000
years ago until about 10,000 years ago. Plant material extracted
from the ground sloth dung indicates that, for much of this time
period, the climate was much cooler than it is today. Research on
packrat middens of similar age in the eastern Grand Canyon also
indicate that this was a much cooler, wetter time, with the shift
toward modern vegetation patterns beginning about 13,000 years
ago. Holdovers from the late Pleistocene, these animals survived the
transition to the warmer post-glacial climate of the Holocene.

The giant Shasta ground sloth,
whose remains were discovered
in the inner canyon in the
1930s.

THE STRUCTURE OF THE LAND

Sinking Ship, visible along the East Rim Drive between Grandview and Moran Points. The gently dipping beds are an expression of the Grandview Monocline.

A great deal of time passed after the Paleozoic rocks at Grand Canyon were deposited and before the Colorado River began to carve its canyon. During that time, this entire part of the country experienced a series of geologic events that gave it much of the architecture and configuration it has today. The most significant of those was the Laramide Orogeny. Unraveling these events is within the realm of structural geology. Unlike the stratigrapher concerned with the characteristics of individual rock layers, the structural geologist is concerned with what has since happened to these bodies of rock, the degree to which they have been deformed or displaced, and their relationship to one another. This information provides important clues to the geologic history of a place, particularly for those intervals of time for which a more detailed sedimentary record is not available. It also adds to our understanding of the look of the land and the origin of the landscape.

In some parts of the country — most of the American Midwest, for example, where great thicknesses of flat-lying beds of sedimentary rock lie only a few hundred feet above sea level — geologic structure tends to be simple and straightforward. In the more mountainous regions of North America, much of the American West in fact, geologic structure can be complex because igneous, metamorphic, and sedimentary rocks are typically folded, broken, crumpled, uplifted, and the pieces jumbled almost beyond recognition.

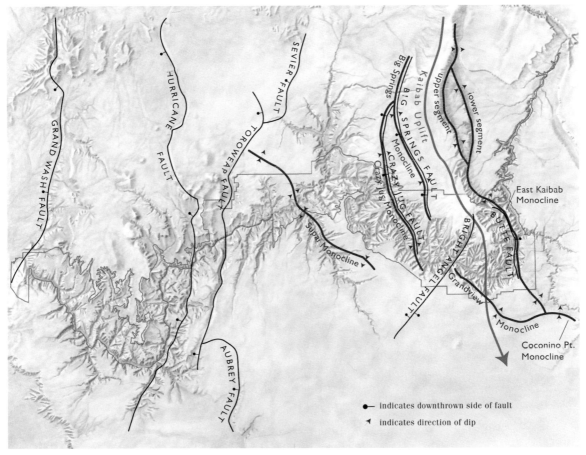

Major structural features of the Grand Canyon region.

The rocks of the Grand Canyon region, and of the Colorado Plateau in general, lie somewhere in between these two extremes. While much of this region has been uplifted thousands of feet above sea level, the processes have occurred in a way that preserved much of the region's original configuration. Nonetheless, large

FAULT TYPES

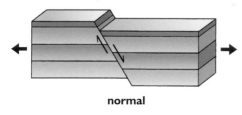

normal

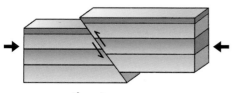

thrust or reverse

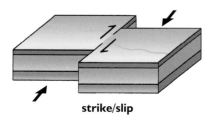

strike/slip

structural features — folds, faults, and uplifts, to name a few — are evident and offer clues to the origin of this landscape.

A structural map of Grand Canyon *(on page 45)* identifies distinct structural features. These in turn reflect the larger forces at work within the earth's crust. Grand Canyon Village rests on the south flank of the structural feature known as the Kaibab Uplift, a portion of the Colorado Plateau that has been lifted a few thousand feet above the surrounding area. The Kaibab Uplift is an irregular, elongate dome. The Colorado River cuts across the Kaibab Uplift, not at its highest point, but about a third of the way up its southern slope. That portion of the uplift north of the Colorado River is known as the Kaibab Plateau; to the south it is known as the Coconino Plateau.

The highest portion of the Kaibab Uplift — more than 9,000 feet above sea level — lies about twelve miles north of the canyon's North Rim. From this high point, near the north entrance station to the park, the strata dip gently to the south and somewhat less gently to the north. This accounts for the fact that the North Rim is 1,000 feet higher than the South Rim.

From the crest of the uplift, on the North Rim's Kaibab Plateau, rain and melting snow flow south into the canyon; that which falls on the South Rim's Coconino Plateau flows away from the canyon. Due to the higher elevation, the North Rim also receives twice as much precipitation as the South Rim. For these reasons, erosion north of the Colorado River in this part of the canyon has proceeded more quickly and effectively. The distance from the North Rim to the river is nearly twice that of the distance from the South Rim to the river. All of the big springs in the inner canyon issue from the north wall. Roaring Springs, on the north side of the canyon, supplies all of the water for the South Rim by way of a small trans-canyon pipeline.

If we look north to Utah, we see that such uplifts are very much a part of the structure of the entire Colorado Plateau. Bryce Canyon National Park is located on the edge of the Paunsaugunt Plateau, over 8,000 feet above sea level. High above the new Grand Staircase–Escalante National Monument, the Aquarius Plateau sits 10,000 feet above sea level. Monument Valley is developed astride a major uplift known as the Monument Upwarp. Between these uplifts

Faults play an important role in shaping the canyon, but a less obvious role than one might imagine. Many of the major north-south drainages that cross the canyon follow major faults or fault zones. These fault zones, where the bedrock is fractured, are zones of weakness along which water can readily travel and erosion proceeds more effectively. The Bright Angel Fault is a classic example: Both the Bright Angel Trail and the North Kaibab Trail follow Bright Angel Canyon, which has eroded headward from the Colorado River along the trace of the Bright Angel Fault.

often are found areas of subsidence and deposition known as basins.

Other easily recognized structural features include various kinds of faults and folds. Faults are fractures or breaks in the earth's crust along which movement has occurred. The term may refer to a single plane along which discernible movement has occurred, or to fracture zones made up of dozens of smaller faults whose collective displacement can be quite large. Individual faults can be difficult to recognize, but the offset that occurs along these faults is often striking. Fractures along which little or no movement has occurred are known simply as cracks, or joints, and are the most common and easily recognized structural features.

A number of major faults cross the canyon. These faults tend to run north-south, somewhat parallel to one another. Many of them originated in Precambrian times. Inactive for long periods of time, they were reactivated in the Late Miocene Epoch (Late Tertiary), and many of them are active today. Residents of Grand Canyon Village occasionally feel small earthquakes that are the result of small periodic movements along one of these faults.

Folds are portions of the earth's crust that have been bent or curved, the result of deformation that occurs because of stresses within the crust. We define folds according to their orientation. Folded strata are often highly fractured internally.

Bright Angel Canyon

East Kaibab Monocline

On the Colorado Plateau, one of the most common folds is the monocline, a single-limbed fold, with one flexure in a series of otherwise flat or gently dipping beds. Sinking Ship, visible from Moran Point on the road to Desert View, is an expression of the Grandview Monocline. The East Kaibab Monocline is best seen exiting the park to the east between Desert View and Cameron. Both are close to the eastern edge of the Kaibab Uplift. These monoclines, far older than the canyon itself, overlie older faults in the crust. Similar monoclines exist along the western edge of the Kaibab Uplift. The Echo Cliffs, which parallel Highway 89 from Cameron to Lees Ferry, are expressions of the Echo Cliffs Monocline. These structures are hallmarks of the regional compressive forces that helped elevate the Colorado Plateau and stand in remarkable contrast to the tight folds and highly deformed strata of mountainous regions like the Appalachians of the eastern United States.

The Landscape Today

mesa

butte

Mesas and buttes are erosional landforms common in arid regions with horizontal layers of sedimentary rock. The two landforms are distinguished by the width of their summits, relative to their height.

Landscape is a product of many things: the rocks that form the surface of the land, the degree to which those rocks have been deformed, and climate. Much of the American Midwest is underlain by rocks similar to those of the Grand Canyon region, but the fact that they lie much closer to sea level and have a much wetter climate produces a landscape that is strikingly different. The Grand Canyon region and the Colorado Plateau are characterized by gently deformed, flat-lying sedimentary rocks. The climate — a function of elevation and, hence, rainfall — ranges from desert at the canyon bottom (with less than eight inches of precipitation each year) to the moist, green, subalpine forests of the high plateaus (which receive over thirty inches of precipitation each year). In this dry land far above sea level with its scarcity of thick vegetation and perennial streams, erosion proceeds at a dramatic rate. Although the landscape of the Grand Canyon region and the Colorado Plateau is young, the regional uplift in this part of the country in the past few million years has accentuated the effects of erosion.

The erosional landforms that are a familiar part of the Grand Canyon area reflect all of these things. Mesas and buttes are common erosional landforms in this part of the world; the more resistant beds of sandstone and limestone form caprocks over the softer layers of shale. Elongate ridgelines like the Echo Cliffs occur where the edge of a gently folded monocline intersects the earth's surface.

The configuration of the canyon itself, so much a part of its visual impact, is likewise a reflection of both structure and climate. The diversity of the sedimentary layers exposed in the canyon walls — alternating beds of sandstone, shale, and limestone — account for the canyon's terraced appearance. Cliff faces are associated with thick layers of sandstone or limestone, which in this climate are resistant to erosion. The slopes at the base of these cliffs are

Water is the principle agent of erosion at Grand Canyon. Much of it arrives as heavy rains that accompany afternoon thunderstorms almost daily in the summer months.

At Grand Canyon, and throughout the American Southwest, color is a striking element of the landscape, partly because of the sheer volume and variety of exposed rock. Each layer has its own inherent color, which may be affected by trace amounts of various elements. The characteristic green of the Bright Angel Shale, for instance, is due to the clay mineral glauconite.

The colors we see on the exposed rock face are often different from the color of the rock itself, the result of weathering, the interaction of water and air with the rocks themselves. Iron, for example, weathers to form characteristically rust-colored iron oxide. These mineral stains are sometimes carried by water onto the cliff face below. The iron-rich beds of the Supai Group are the source of the red color we see on the face of the underlying Redwall Limestone.

Another weathering phenomenon in the desert Southwest is desert varnish (right),

a dark coating of iron oxide with trace amounts of manganese oxide and silica that forms on the exposed surface of rocks over time. This complex process involves the migration of mineralized solutions from within the

rock to the exterior, where evaporation allows for the formation of a thin crust. In recent years, researchers have proposed that microorganisms play a role in its formation. This fragile coating takes many years to form.

associated with the softer, more easily eroded shales. The broad expanse of the Tonto Platform that stretches from the base of the Redwall cliff to the edge of the Inner Gorge exists because of the thick exposures of Bright Angel Shale which outcrop at that point. The V-shaped Inner Gorge results from the more resistant igneous and metamorphic rocks exposed there.

In the long run, landscape is not so easily defined. It includes the quality of light, the apparently haphazard configuration of land-forms, the plants and animals that inhabit it, and myriad ineffable qualities. Increasingly, it includes the cultural artifacts, from the sublime to the ridiculous, of our own species.

There is no more stunning landscape than that of the Grand Canyon region, and there are few places that are more revelatory. Those who seek to know more about origin and process at Grand Canyon come away with an understanding of the complex history of our planet in all its grandeur. Our understanding will always be incomplete, but that only enhances our appreciation of what we do know.

Geology is a science that is deeply rooted in aesthetics. One cannot help but imagine that our sense of beauty must have had its origins somewhere in an appreciation of landscape. For some, that sense of beauty is closely tied to a sense of wonder.

The most beautiful and most profound emotion we can experience is the sensation of the mystical. It is the source of all true science.

— Albert Einstein

THE COLORADO RIVER

In the remote western portion of the park, Grand Canyon has a very different profile. From this overlook at the foot of the Toroweap Valley it's a drop of 3,000 vertical feet from the rim to the Colorado River.

Grand Canyon is part of the greater Colorado River system, which stretches more than 1,400 miles from its source in the Rocky Mountains of Colorado to its mouth at the Gulf of California. The river's headwaters lie at an altitude of over 14,000 feet; much of the water that flows through the canyon is snowmelt. The Colorado River and its tributaries drain an area of 244,000 square miles. Its major tributary is the Green River, which originates in the Rocky Mountains of Wyoming and contributes nearly as much water as the Colorado River above their confluence. The Colorado in its present course crosses three geologic provinces: the Rocky Mountains, the

Colorado Plateau, and the Basin and Range. The Colorado River is responsible for the canyon's existence, and the role it has played in carving the canyon has much to do with the erosive power of running water in an arid land.

The Dynamics of Rivers

How does a river like the Colorado carve a canyon like the Grand? Not by itself. Running water has the ability to do two things: carry sediment (gravel, sand, silt, clay, or some combination of the four) and carve a channel in the rock through which it flows, be it a canyon, valley, or broad alluvial plain. The degree to which a stream does these things depends upon several factors: how fast the water is moving, how much water there is, the periodicity of its flow, the slope of the stream bed, etc. In general terms, we talk about the energy of a given stream or how much power it has to get from its source to its base level, which for most water is the sea. Those rivers that originate at high altitudes have a great deal of energy. The fact that Grand Canyon exists on the Colorado Plateau is not coincidental. Here in this high, arid land, thousands of feet above sea level, all running water has an increased capacity for erosion and the ability to create spectacular erosional features, among them deeply incised canyons.

If we classify rivers according to the amount of water they carry, the Colorado River ranks twenty-seventh in North America. It is the gradient, the slope of its streambed, that determines how fast the water is moving, the nature of the debris it carries, and, ultimately, its erosive power.

The longest and best-known river in North America is the Mississippi. It flows for a distance of 2,340 miles from its headwaters in Minnesota to its mouth at the Gulf of Mexico. In that distance, it drops just about a thousand feet in elevation. Though it drains more than a million square miles of moist and humid land and carries a huge amount of water, it is a slow, lazy river, a low-gradient stream dropping less than a foot per mile. The Mississippi carries a great deal of fine-grained sediment to its mouth and deposits much of that mud in broad floodplains along the way. It directs its energy sideways, depositing and reworking the broad

The Colorado River drains an area of 244,000 square miles.

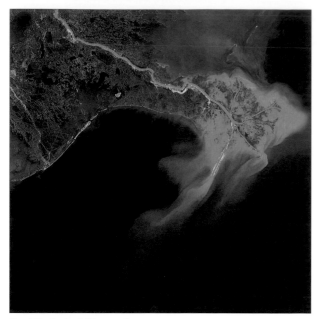

The Mississippi River, a low-gradient steam, drains about a third of the continental U.S. and delivers about 500 million tons of sediment to the Gulf of Mexico each year.

alluvial plains that characterize the Mississippi River Valley.

The Colorado River, by comparison, is half that length, carries less than a tenth the water, and drains a much smaller area. But in its 1,400-mile course, it drops over 14,000 feet, ten feet for every mile it flows. In the 277-mile length of Grand Canyon alone, it drops 2,000 feet. Such high-gradient streams tend to carve narrow, deep canyons rather than broad flood plains.

In a few million years as the Colorado River carved its canyon, it carried a huge amount of sediment downstream. In pre-dam years (prior to 1963) the Colorado River carried about 380,000 tons of sediment through the canyon on an average day. We estimate that more than 1,000 cubic miles of debris has been removed from Grand Canyon by erosion.

Geologists have discovered — sometimes the hard way — that much of a river's work is accomplished during times of flood, brief periods during which the flow is many times its average. In those pre-dam years the flow of the Colorado fluctuated seasonally from 700 cubic feet per second (cfs) to 100,000 cfs. There is evidence of regular flows well above 100,000 cfs, and we have record of one late-nineteenth-century flood as high as 300,000 cfs. We can only try to imagine the power of the Colorado River during such times.

The Colorado River is not the only stream in the canyon that floods. Its major tributaries, the Paria and the Little Colorado Rivers, also carry large volumes of water and sediment during times of heavy rain or melting snow. The hundreds of steep-walled side canyons that feed into the main course of the river throughout the canyon are just as important in shaping the canyon.

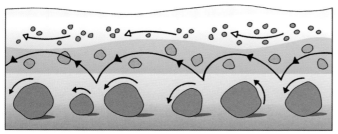

A river transports material several ways. Larger pieces of rock are moved along the bottom; finer particles of sand and silt are carried in suspension. All of this varies with the speed and turbulence of the water.

The Shape of the Canyon

The river is responsible for the depth of the canyon, and it is the primary agent of transport for all of the sediment that has been removed, but other erosional processes have given the canyon its fantastic shapes and forms so visually striking to all who stand gazing from the rim.

The work that a river does, it does at river level. The shape of the canyon above the high-water mark is the result of erosion by water from other sources. In the Grand Canyon region rain tends to come intermittently in heavy thunderstorms during the summer months. Winter snows accumulate to a depth of up to twelve feet on the Kaibab Plateau north of the Colorado River. Most of this water eventually finds its way to the river, and in doing so it shapes the walls of the canyon, carrying debris and house-sized boulders as it goes.

Most of the narrow, steep tributary canyons that feed the main channel carry water only during times of rainfall and snowmelt. From the rim we see only dry washes strewn with boulders and a few trees. But when these streams run they run with a fury, working their way back into the plateau through headward erosion and contributing an enormous amount of rocky debris to the river.

Flash flood in Blue Mountain Canyon, a tributary to Diamond Creek. Floods like this one transport an enormous amount of rock.

Usually the debris which accumulates at the mouth of tributary canyons is ultimately carried away by even larger floods along the Colorado River itself, floods that course through the canyon regularly if not often. One concern today is that, in the absence of such floods, due to the controlled flows that emerge from Glen Canyon Dam upstream, the canyon may eventually become choked with debris.

The Origin of Rapids

Rapids within Grand Canyon occur where the river channel is choked with erosional debris, usually at the mouths of side canyons, forcing the constricted flow of the river to increase in speed and turbulence. Upstream from a rapid, the river achieves a kind of flat-water calm, as if it were pooling above a dam. Then in the course of a few feet, the level of the river can drop ten, twenty feet or more.

The complex hydraulics of such flow creates whirlpools, eddies, and holes. Below the rapid, the river once again becomes more tranquil.

Of the 2,000-foot loss in elevation between Lees Ferry and the Grand Wash Cliffs, about half occurs in rapids, most of which exist where steep side canyons and tributary drainages intersect the river. These drainages tend to develop along regional fault lines and fracture zones, where erosion proceeds rapidly and effectively. Low-gradient tributaries, of which there are few, do not create such constrictions. The entrance of the Paria River at Lees Ferry creates a small riffle, its flow visibly running side-by-side with that of the Colorado for awhile until the two mingle downstream. Steep, high-gradient tributaries contribute a great deal of debris that tends to arrive at the river not in a slow, steady stream but during times of flood. Crystal Rapid, considered one of the major boating challenges on the Colorado River today, was created in 1966 when the North Rim received an overnight rain of fourteen inches, much of which rushed down Crystal Canyon, carrying with it an almost unimaginable amount of material in a debris flow.

Nearly all of the rapids on the Colorado River within Grand Canyon are the result of debris from tributary canyons that constrict the channel of the river.

Debris Flows

Some of the rock debris which finds its way to the river from these tributary canyons does so in times of moderate flood, as sediment-laden streams work their way to the river, undermining large boulders as they go. Occasionally, however, an interesting phenomenon known as a debris flow occurs. Debris flows are moving masses of rock fragments, soil, mud, and water. They occur when heavy rains saturate the ground, and they carry an immense quantity of rock debris to the river in a very brief period of time. Containing 10 to 40 percent water by volume, much less than that in floodwaters, they can achieve speeds of up to one hundred miles per hour. They are the primary method of sediment transport in the small, steep drainages that feed the Colorado River in Grand Canyon. Few people have ever seen a debris flow, but they have been witnessed

March 6, 1996, debris flow at the mouth of Prospect Canyon. Debris from Prospect Canyon is responsible for Lava Falls, one of the most challenging stretches of whitewater in the canyon.

and captured on film. An average of two debris flows reach the river each year, enough over the course of the canyon's history to reshape many rapids. Debris flows are another reminder that, although we tend to think of geologic processes as operating at a slow and steady pace, this is often not the case.

The History of the Colorado River

Unraveling the detailed history of the canyon and river is perhaps the most difficult task geologists face at Grand Canyon. The history of Grand Canyon is the history of the Colorado River. And there's the challenge, for much of the evidence we seek about the details of its age, history, and configuration is long gone, carried downstream by the river itself, or lost to us as a result of on-going erosional processes. The evidence we do find is scarce, fragmented, and enigmatic. There are a few givens, however.

We know that the Colorado River cannot always have existed in its present form. The Rocky Mountains have been around for 60 million years, and the earliest drainage systems that formed on their western slopes — ancestors of the present-day Colorado River system — are therefore as old. But today the Colorado River finds an outlet at the Gulf of California, and the Gulf of California is no older than 5.5 million years. Furthermore, there is compelling evidence that the lower Colorado River below Grand Canyon is younger than 5 million years, and in the western Grand Canyon there is evidence that, by a million years ago, that portion of the canyon was virtually the size it is today. This means that Grand Canyon must have been carved within a period of as little as 3 to 5 million years.

Scientists then realized that different portions of the Colorado River are of different ages. Remember that the river crosses three geologic provinces in its course, and distinct portions of what we now call the Colorado River have very dissimilar histories. It is possible, for instance, that the upper Colorado River, which has drained the Rocky Mountains since their birth, was for many years a river without a mouth, an internal drainage system that fed a series of lakes in the continental interior, with no outlet to the sea. This is true today of a number of streams in the Great Basin.

Today there are fifteen major reservoirs along the Colorado River and its tributaries, and the flow of water from beginning to end is carefully controlled by a series of dams. The water is spoken for before it ever reaches Grand Canyon, carefully doled out to each of seven western states and Mexico by the Colorado River Compact of 1922, and by international treaty. The two dams and reservoirs that exert the greatest influence on Grand Canyon are Glen Canyon Dam–Lake Powell at the upper end and Hoover Dam–Lake Mead at the lower end. Lake Mead has flooded portions of the lower Grand Canyon since about 1940, and Glen Canyon Dam *(below)*, fifteen miles above Lees Ferry, has carefully regulated the flow of water through Grand Canyon since the dam was completed in 1963.

It is likely that the river we know today as the Colorado came into existence through the integration of several smaller river systems. The most popular theory holds that a number of events occurred to create a free-flowing Colorado River from its headwaters to the Gulf of California. First was the opening of the Gulf of California, about 5.5 million years ago, when movement along the San Andreas Fault created a major depression in that part of North America. Then from the gulf, a "lower" Colorado River system began to erode headward up to the western edge of the Colorado Plateau. Somewhere west of the Kaibab Uplift, it eroded into and captured the land-locked upper Colorado River, giving it an outlet to the gulf.

The precise times at which these events likely occurred is difficult to know, but we do know that by the beginning of the Pliocene, about 5 million years ago, the through-flowing Colorado River had come into being.

All the while, the uplift of the Colorado Plateau — which began with the Laramide Orogeny at the end of the Mesozoic — continued. By some estimates, there has been as much as 1,000 feet of uplift just in the past million years. This would explain not only the extraordinary depth of Grand Canyon, but also the prevalence throughout the Colorado Plateau of canyons that are much deeper than they are wide. Evidence of this fairly recent pulse of uplift is all around us. Witness the "entrenched" meanders of the San Juan River, a tributary of the Colorado just to the north. Geologists have long held that such features are a clear indication of the sudden onset of gradual but persistent regional uplift within a short span of geologic time. In a part of the world where all streams tend to carve deep canyons, is it such a surprise that the mightiest such stream — the Colorado River — has carved the most spectacular canyon of them all?

We may never have all of the details and may never know the precise history of the Colorado River. But we know the kinds of events that must have occurred. We know much about the process, and the age of the canyon — at least within a narrow range — has been reasonably well established. It is ironic that, in some respects, we know more about the Paleozoic history of this region of the continent than we do about the last 10 million years, but this is a land where erosion has prevailed for more than 200 million years. Geologists, used to the rigors of a discipline where years of investigation yield an incomplete picture at best, are not discouraged by the prospect, and Grand Canyon is a very worthwhile place in which to seek the answers.

Civilization exists by geological consent, subject to change without notice.

— Will Durant

GLOSSARY

basalt — a dark-colored volcanic rock, rich in iron and magnesium.

butte — a flat-topped erosional landform common in arid regions of flat-lying sedimentary or volcanic rock. It is differentiated from a mesa in that it is taller than it is wide.

calcite — a common rock-forming mineral composed of calcium carbonate.

Colorado Plateau — the geologic province bounded on the west and south by the Basin and Range province and on the east and north by the Rocky Mountains, characterized by thick sequences of flat-lying sedimentary rocks that have been uplifted thousands of feet above sea level. Includes portions of Utah, Colorado, New Mexico, and Arizona.

cross-bedding — a characteristic of sedimentary rocks in which internal layers have been deposited at an angle to the predominant bedding orientation.

debris flow — a moving mass of rock fragments, mud, and water. Debris flows contain up to 40 percent water by volume. Many flash floods are in fact debris flows.

dike — a veinlike igneous intrusion that cuts across the grain of the rock into which it is intruded. At Grand Canyon these are generally light-colored; they are clearly visible in the walls of the Inner Gorge.

erosion — the general term for those processes that wear away and carry off materials on the earth's surface. Includes physical and chemical weathering and the processes which transport these rock materials (wind, water, etc.).

extrusive — the general term for igneous rocks that erupt and cool at the earth's surface. All volcanic rocks are extrusive.

fault — a fracture in the earth's surface along which movement has occurred.

formation — the formal name geologists assign to a rock unit that is recognizable, easily differentiated from adjacent strata, and traceable for some distance — the Kaibab Formation, for example.

gneiss — a common metamorphic rock, similar in composition to granite but in which the mineral grains are aligned in distinct bands.

granite — a common intrusive igneous rock, composed primarily of quartz and feldspar.

Great Unconformity — the name John Wesley Powell assigned to the immense gap in the geologic record at Grand Canyon between the igneous and metamorphic rocks of the Inner Gorge and the overlying Paleozoic rocks.

headward erosion — the process whereby a stream extends its reach in an upstream direction.

igneous — referring to a rock that forms when molten rock cools, either slowly at depth (intrusive) or more quickly at the earth's surface (extrusive).

intrusive — the general term for igneous rocks that cool slowly beneath the surface of the earth — granite, for example.

isotope — variety of an element that differs in the number of neutrons it possesses in its atomic nuclei. Unstable (radioactive) isotopes generally decay at a predictable and measurable rate, allowing for the development of a radiometric time scale.

lava — the general term for molten rock that erupts and cools at the earth's surface.

limestone — a common sedimentary rock, generally of biologic origin, composed primarily of calcium carbonate.

lithospheric plates — the name given to the large, rigid but mobile pieces of the earth's crust that migrate relative to one another across the surface of the earth. They can be oceanic, continental, or both. The interaction of lithospheric plates at plate boundaries is responsible for much of the geologic activity on our planet.

marine — the general term for rocks that accumulate in salt water (ocean) environments, rather than freshwater environments (lakes, rivers, and streams).

mesa — a flat-topped erosional landform common in arid regions of flat-lying sedimentary or volcanic rock. It is differentiated from a butte in that it is generally wider than it is tall.

metamorphic — any rock that forms from the alteration, through heat and pressure over time, of a pre-existing rock.

monocline — a single fold or bend in the earth's crust. Many monoclines overlie faults deep beneath the earth's surface.

orogeny — the term used to describe those geologic events responsible for the folding, faulting, uplift, and deformation of mountainous regions. Orogenies represent a period of geologic activity rather than a single event.

plate tectonics — the widely accepted theory that the surface of the earth is composed of rigid but mobile lithospheric plates that move and interact with one another, causing much of the geologic activity on the planet.

pluton — a body of igneous rock that forms when molten rock rises from the earth's interior into the overlying crust and cools without erupting at the earth's surface.

regression — the event in which ocean waters retreat from large portions of a continental land mass, exposing it to erosion.

sandstone — a common sedimentary rock composed primarily of particles of sand, with minor amounts of silt and clay. These particles are cemented together, generally by trace amounts of calcite or silica.

sedimentary — referring to rocks that result from the accumulation and consolidation in layers of particles of rock debris. The formation of sedimentary rocks can involve the transportation of inorganic material like sand or silt (as in sandstone or shale) or the precipitation of organic material (as in limestone).

shale — a common sedimentary rock composed primarily of silt and clay.

silica — quartz (silicon dioxide), one of the most common rock-forming minerals.

transgression — the event in which ocean waters invade large portions of a continental land mass, creating shallow intercontinental seas.

unconformity — a gap in the geologic record, where one rock layer overlies another of substantially greater age. The gap may be due to non-deposition or to erosion of previously existing layers, or both.

weathering — general term for those processes, both physical and chemical, whereby rock materials are decomposed through exposure to elements of the earth's atmosphere.

SUGGESTED READING

The body of literature on Grand Canyon geology is large. The following books are the logical next step for those who would like to know more about some aspect of this story.

Walter Alvarez, *T. rex and the Crater of Doom,* 1997, Princeton University Press.

Arizona Geological Society and the Arizona Geological Survey, *Geological Highway Map of Arizona,* 1998.

Stanley Beus & Michael Morales, *Grand Canyon Geology,* 1990, Oxford University Press.

Steven Carothers & Bryan Brown, *The Colorado River Through Grand Canyon,* 1991, University of Arizona Press.

Wendell Duffield, *Volcanoes of Northern Arizona,* 1998, Grand Canyon Association.

Stephen Jay Gould, *Time's Arrow, Time's Cycle,* 1988, Harvard University Press.

Ivo Lucchitta, *Canyon Maker,* 1988, Museum of Northern Arizona.

Dale Nations & Edmund Stump, *Geology of Arizona,* second edition, 1997, Kendall/Hunt Publishing Company.

Stephen Pyne, *How the Canyon Became Grand, A Short History,* 1998, Viking Press.

PHOTOGRAPHY CREDITS

ILLUSTRATION CREDITS

INDEX

ABOUT THE AUTHOR

Greer Price received his graduate degree in geology
from Washington University in St. Louis. He worked
for seven years as a professional geologist, and
spent ten years with the National Park Service, most
of it at Grand Canyon. His career has involved
teaching, writing, and field work throughout much of
North America. He is currently senior geologist/chief
editor at the New Mexico Bureau of Geology and
Mineral Resources in Socorro, New Mexico.